AF266705

4th Grade Math
Volume 6

© 2013 OnBoard Academics, Inc
Newburyport, MA 01950
800-596-3175
www.onboardacademics.com

ISBN: 978-1-939796-87-5

Table of Contents

Find the Average

Key Vocabulary

average

mean average

line plot

Average

Find the average number of "awesome stars" awarded.
Move the stars among the "awesome" students to find the average.

Share the stars equally among the 4 students.

Finding the Average.

Instead of moving stars among the awesome students you can also find the average
this way.

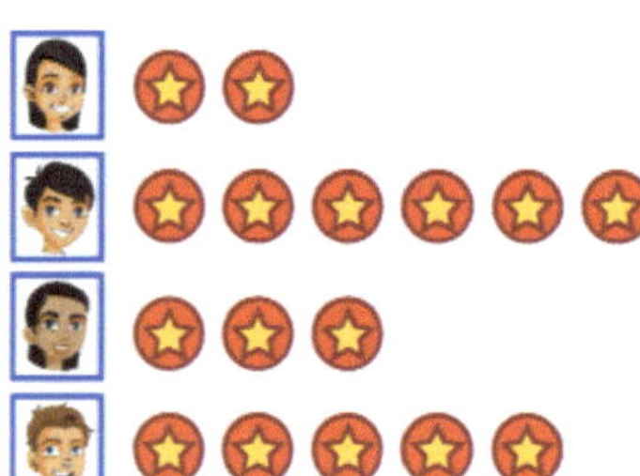

Step 1

Total number of "awesome" stars

16

Step 2

Number of students

4

Step 3

$$\frac{16}{4} = 4$$

Average number of "awesome" stars

     www.onboardacademics.com   6

Find the average number of pets.

Find the average for each group.

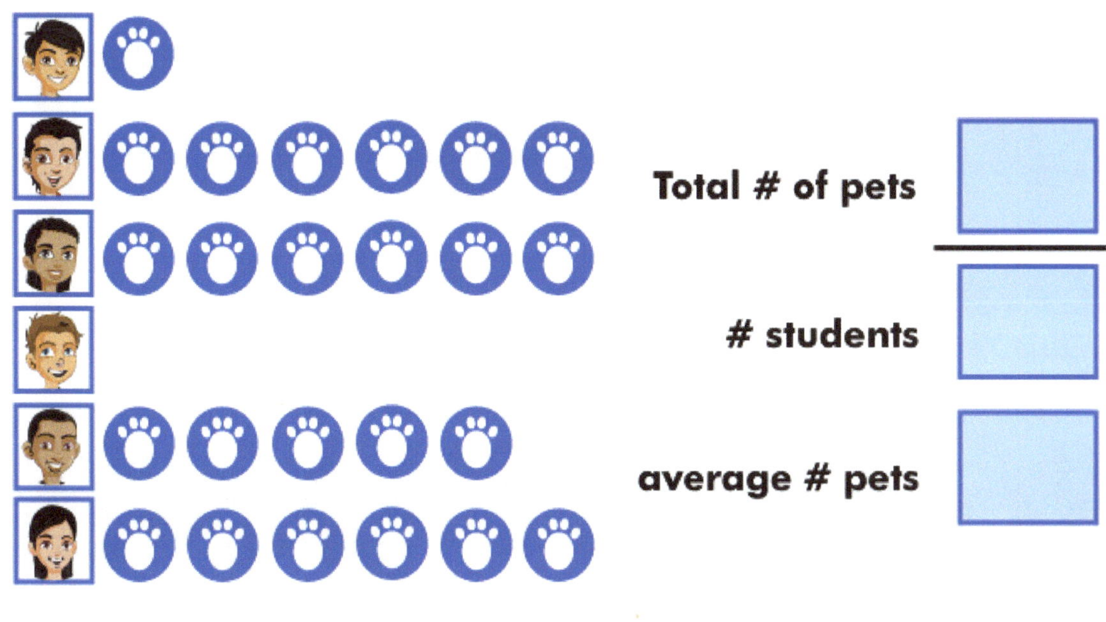

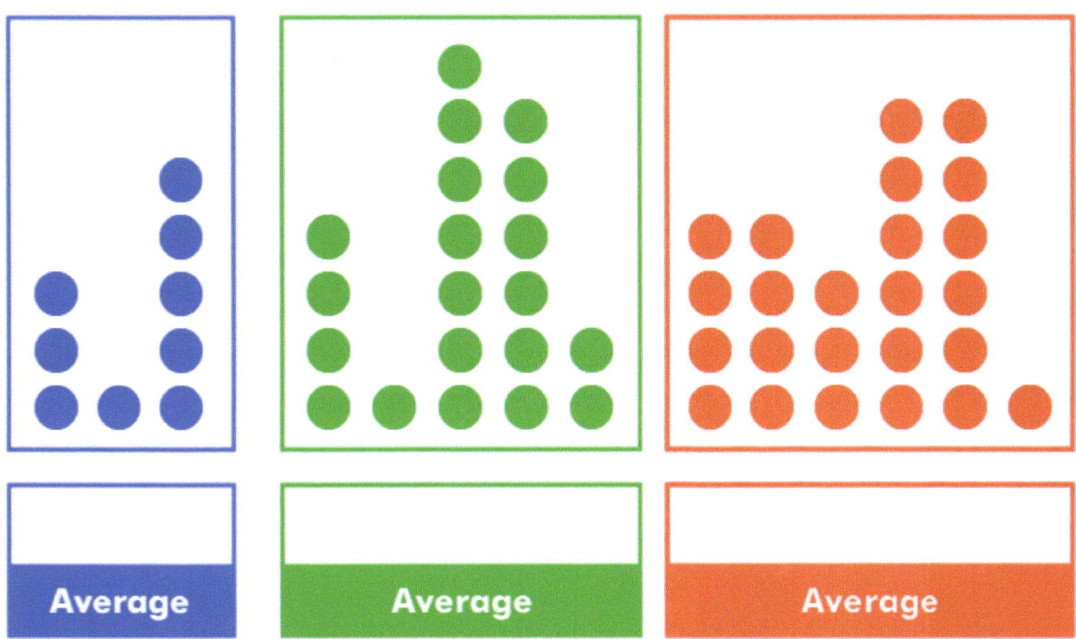

Find the average.

1   Fernando is paid $12 on Saturday, $14 on Sunday, and $4 on Monday. What is his average daily pay?

$

2   In 5 science quizzes, Mia scores 5, 8, 7, 8 and 12. What is her average score?

3   For a math test, Fernando studies for 3 hours, Tony studies for 9 hours, and Jack doesn't study. What is the average number hours spent studying?

Work backwards to find the average.
Move the circles between sisters. Each circle represents one year.

The average age of 3 sisters is 5 years. Alison is 2 years old, Alicia is 9 years old. How old is Tori?

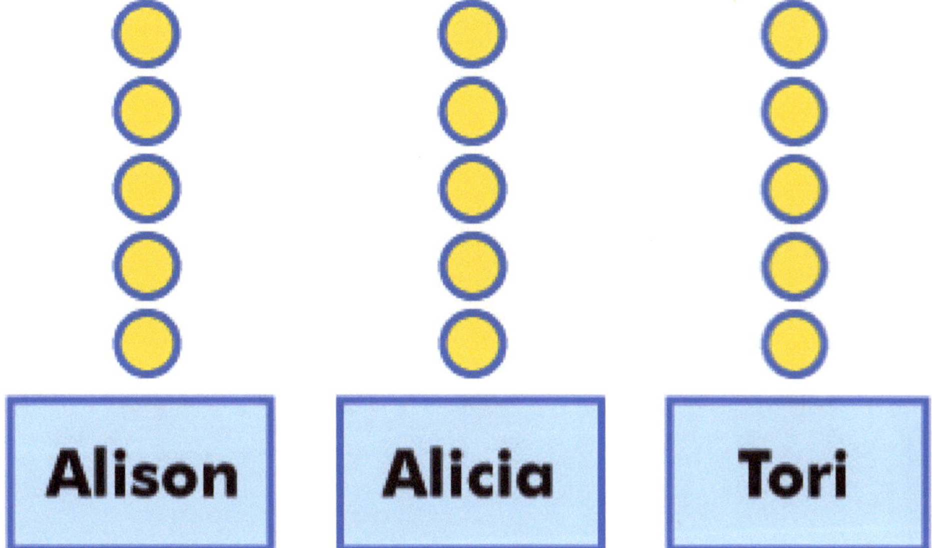

Name___________________________________

Find the Average Quiz

1 **True or false? The average of these three numbers is 7.**
3, 5, 13

2 **What is the average of the data set below?**

A 5

B 7

C 8

D 9

8, 8, 3, 7, 9

3 **The average of 3 numbers is 10. Find the value of n.**
n, 8, 16

4 **If Joe travels 360 miles in 6 hours, what is his average speed per hour (in mph)?**

Measure of Central Tendency

Key Vocabulary

mean

median

mode

range

line plot

Survey of Class 4B
Can you complete this chart?

Number of Children in Family — Class 4B

# Children	Tally	Frequency
1			1
2						4
3	|||	?
4				2
5		0
6		?
7			1

Complete the plot line for the class 4B survey.

# Children	Frequency
1	1
2	4
3	3
4	2
5	0
6	0
7	1

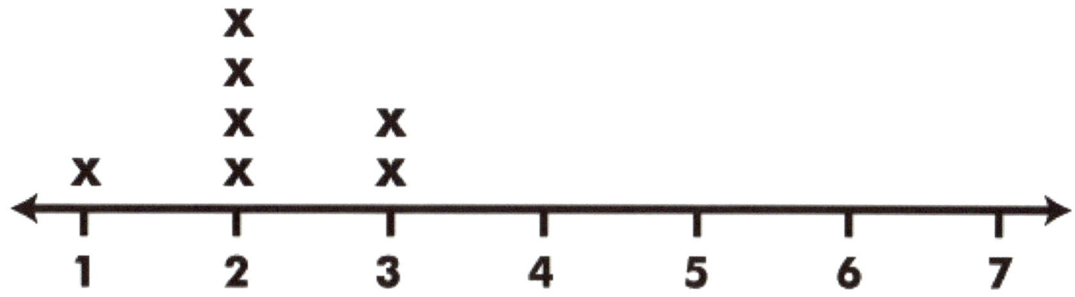

Number of Children in Family — Class 4B

Describe this plot line.

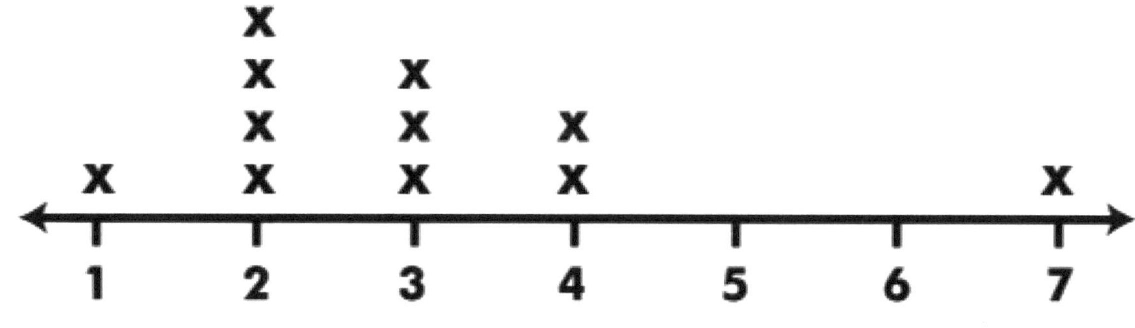

____

____

____

____

____

     www.onboardacademics.com

This question's answers are set up as a subtraction problem to help you find the range. Answer the questions, complete the subtraction to find the range.

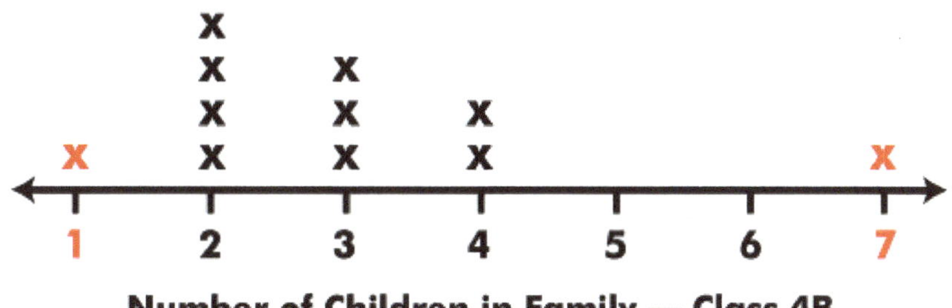

What is the greatest number of children in a family?

What is the least number of children in a family? —

What is the range?

Finding the Mode

Mode means most common.

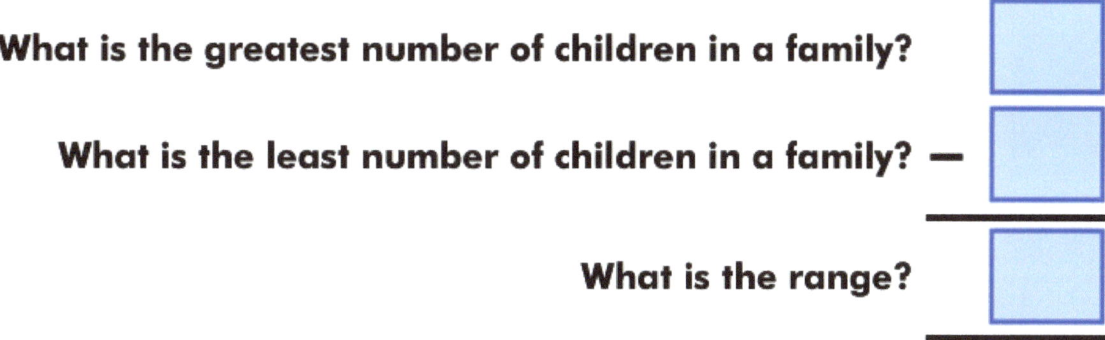

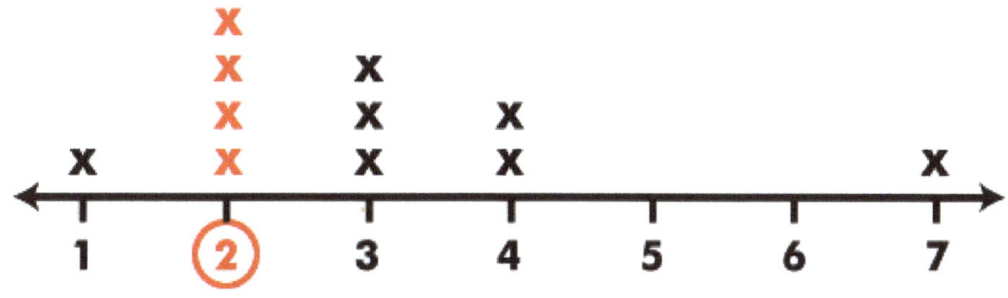

What is the mode?

Find the Median

1, 2, 2, 2, 2, 3, 3, 3, 4, 4, 7

What is the median number of children in a family?

____

Finding the Mean

1 + 8 + 9 + 8 + 0 + 0 + 7 = 33 children

1 + 4 + 3 + 2 + 0 + 0 + 1 = 11 families

Total number of children in all families?

Number of families/students in Class 4B?

Mean number of children in a family?

Name_________________________________

Measure of Central Tendency Quiz

(1) **True or false? The range of this data set is 9.**

(2) **What is the mean of this data set?**

A 4

B 5

C 6

D 4.5

> **1, 2, 3, 4, 5, 5, 8**

(3) **What is the mode of this data set?**

(4) **What is the median of this data set?**

Bar Graphs

Key Vocabulary

bar graphs

double bar graphs

Grade 4 Music Survey

Refer to the bar graph to answer
the questions.

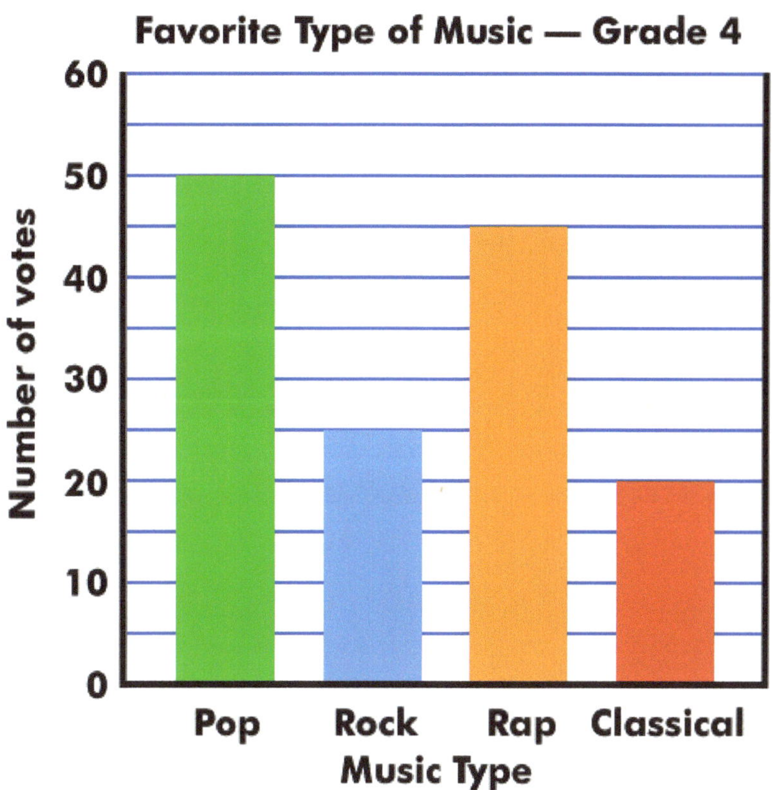

**Which type of music
is the most popular?**

**How many students
voted for this type?**

**How many students
voted for rap?**

**How many students
took part in the
survey?**

Favorite Food survey

The food survey yielded these results.

Favorite Cuisine	# Students
Italian	12
Mexican	7
Chinese	8
Other	2

Use the information from the survey and the elements for the bar graph to complete the bar graph to represent the results of the survey.

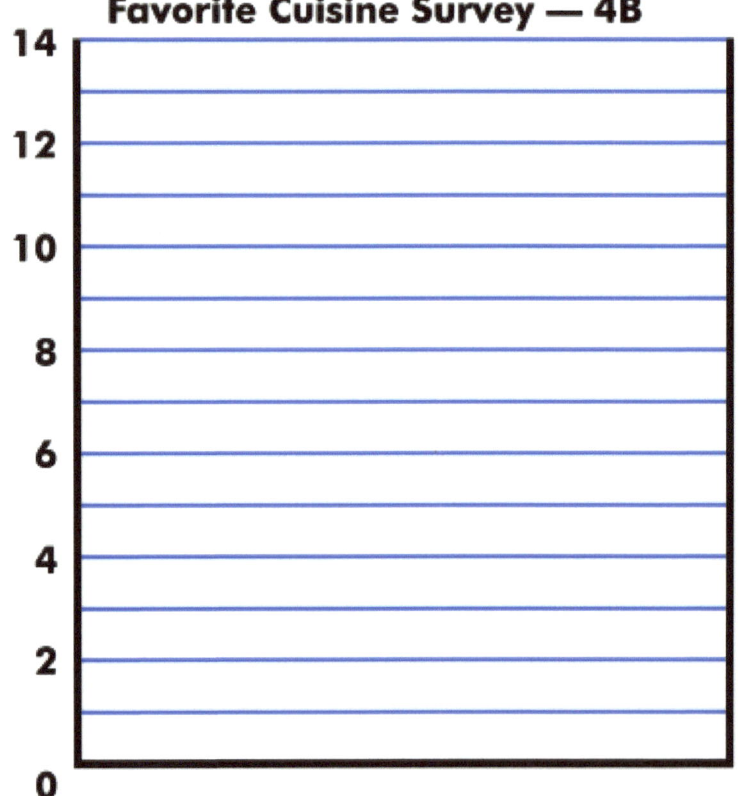

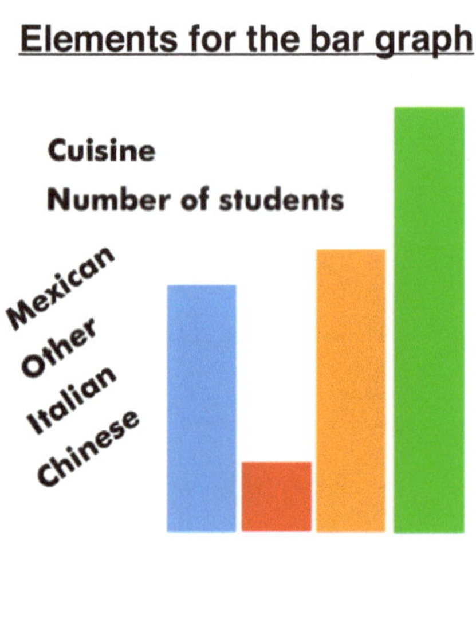

Parts of a Bar Graph

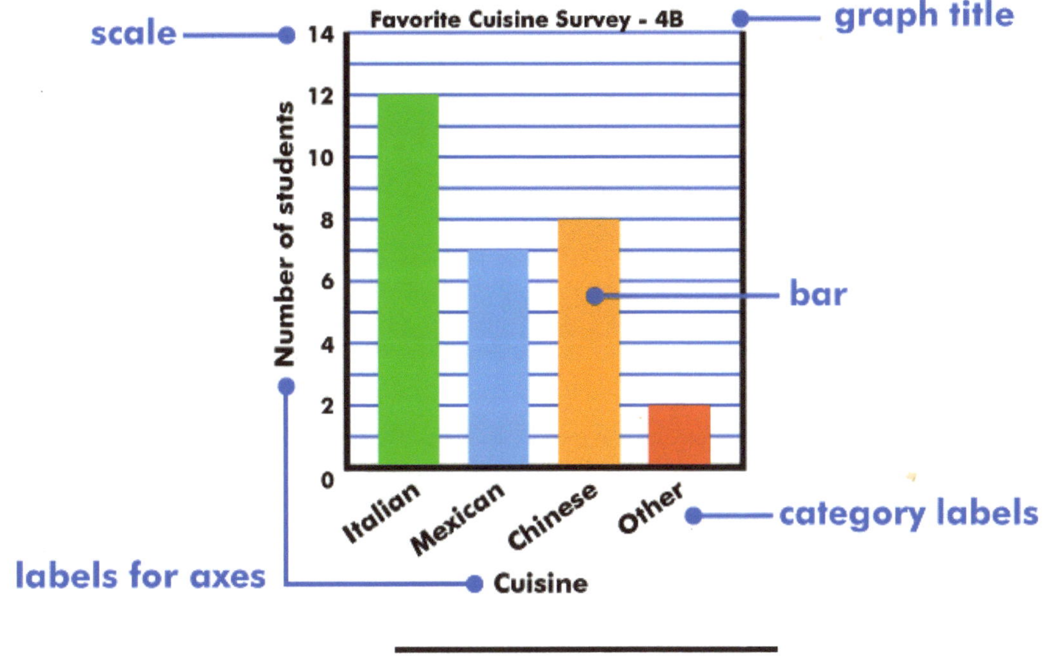

Which ski resort would you choose?

Why?

____

____

____

____

____

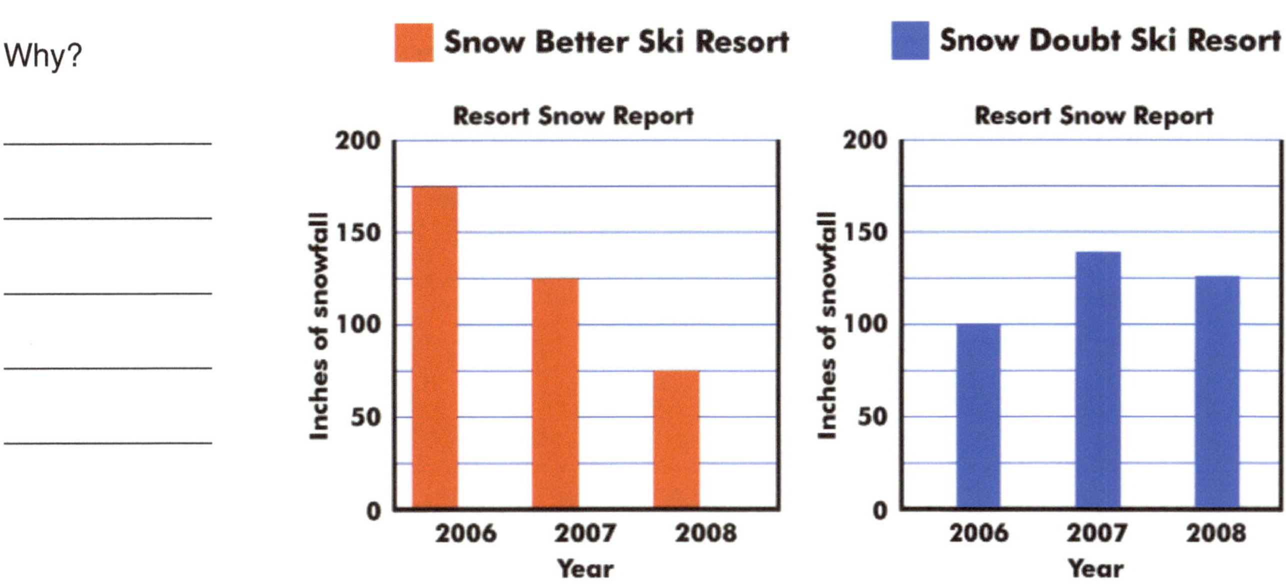

Analyzing Bar Graphs

When you analyze a bar graph, you need to ask questions. Let's think about the last question regarding the best ski resort. Let's suppose you wanted to pick the best resort based on which gets the most snow because you like to ski!

Start by putting the data from both bar graphs on the same bar graph so you can easily see the information.

Then analyze the data by asking questions.

In which years did Snow Better get less snow than Snow Doubt?

How much snow did Snow Better get in 2008?

About how much snow did Snow Doubt get in 2007?

Which resort had the most snow in total in 2006-2008?

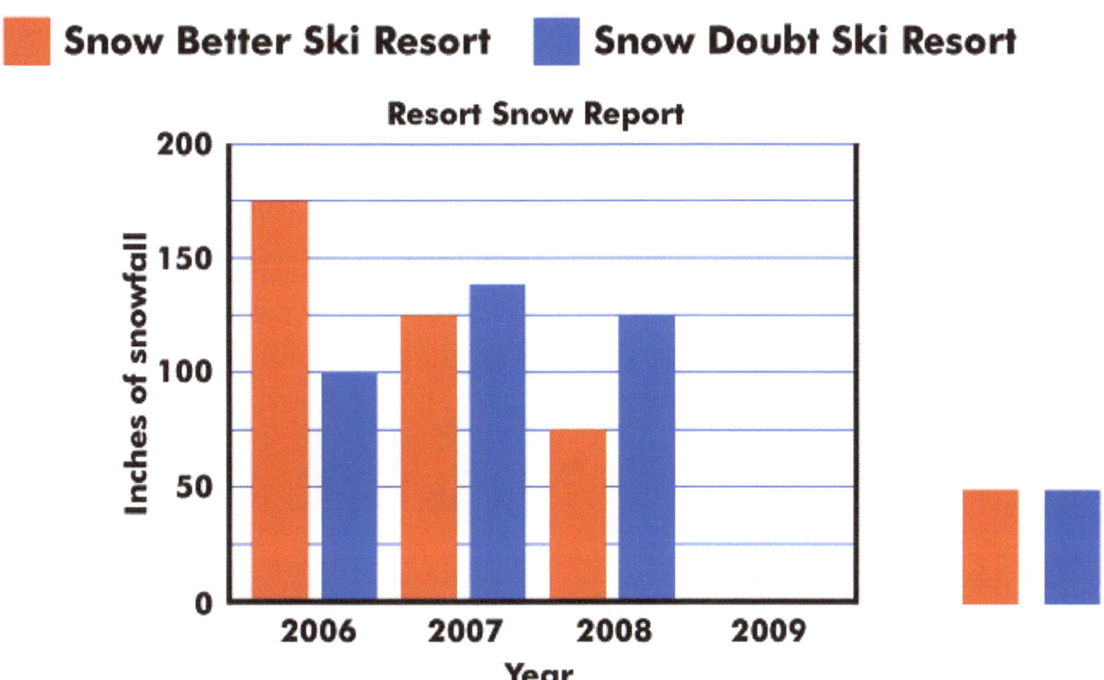

Snow Better received 200 inches of snow during the 2009 season. Snow Doubt received 100 inches. Update the graph.

Name________________________________

Bar Graphs Quiz

1 **True or false? 20 students have blonde or red hair.**

2 **How many students took part in this survey?**

A 50

B 55

C 60

D 65

3 **How many students have brown or black hair?**

4 **How many students *do not* have brown or red hair?**

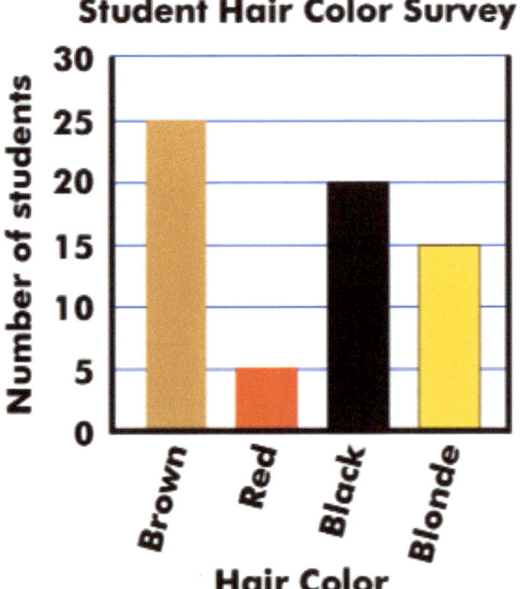

Probability

Key Vocabulary

probability

event

outcome

favorable outcome

equally likely

Which outcomes are equally likely?

Place a checkmark next to the items that you believe will have an equally likely outcome and and X if not.

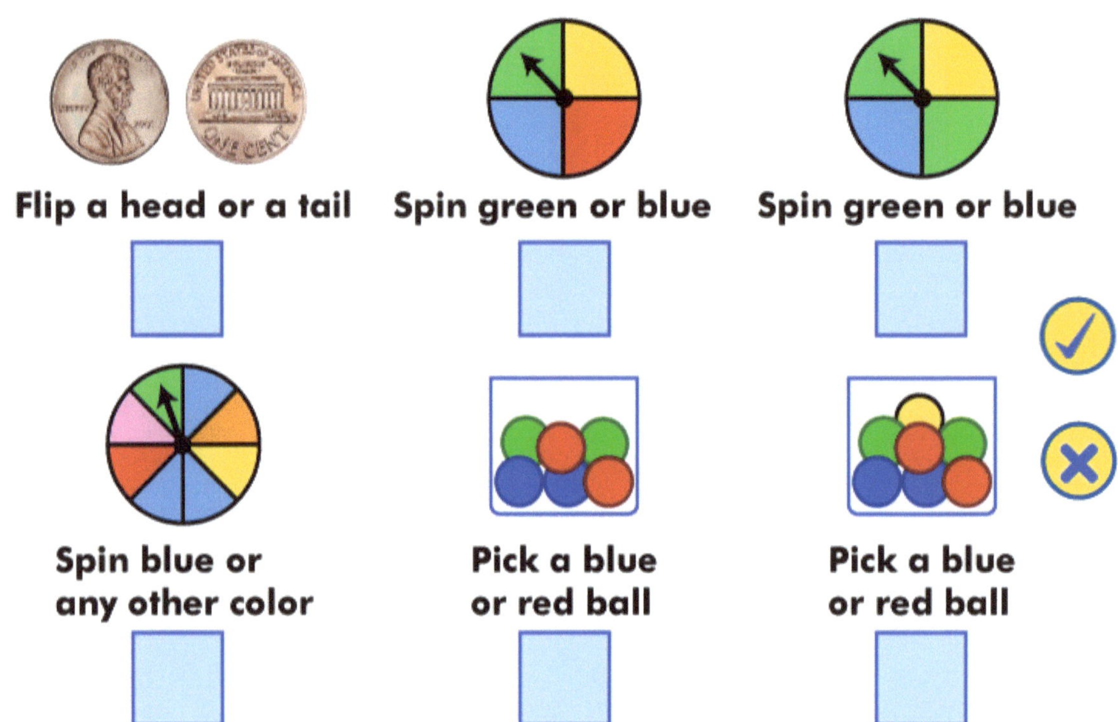

Where do these outcomes belong on the probability line?

Draw a line from the illustration to the proper position on the probability line. One is done for you as an example.

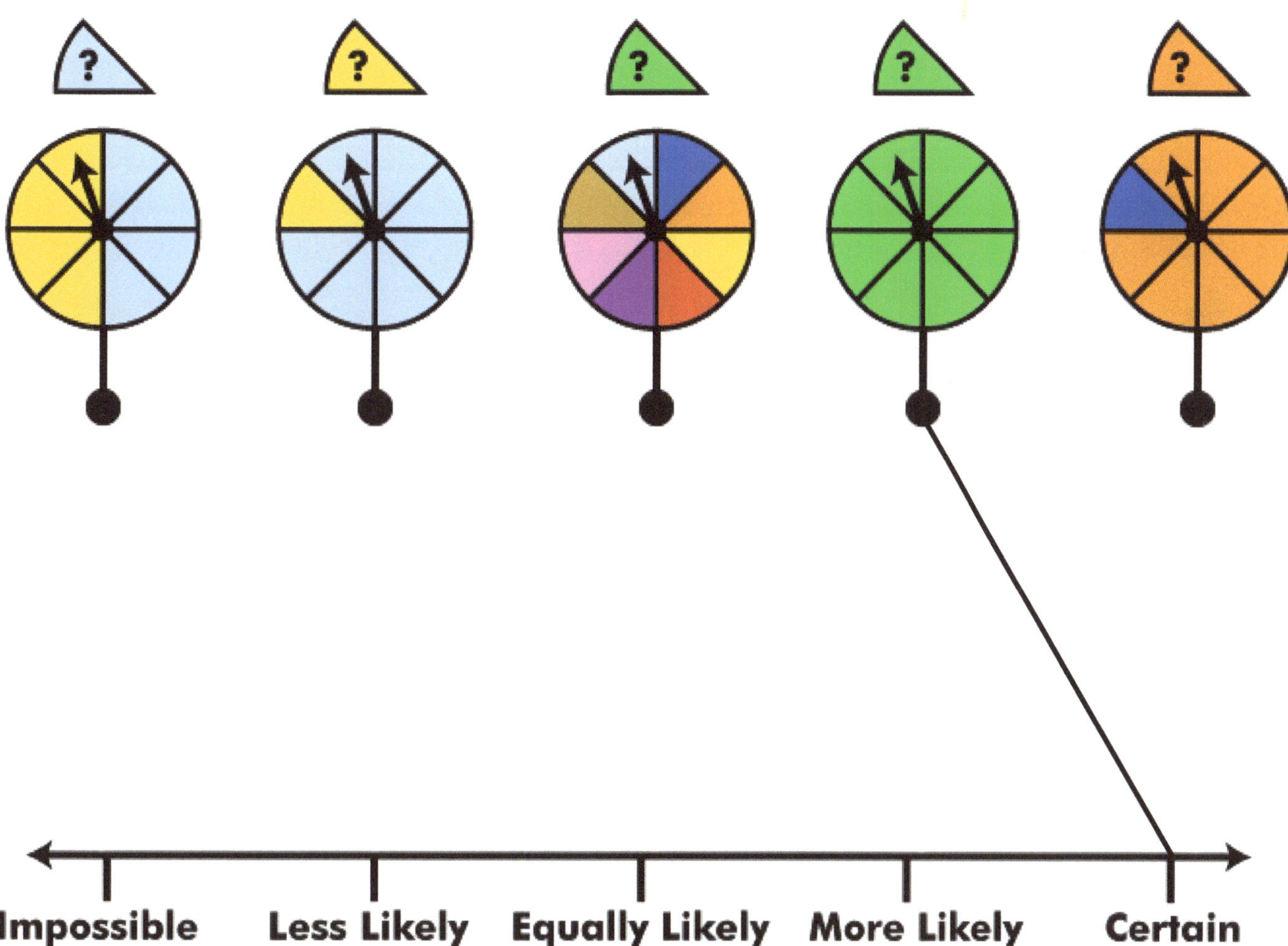

What is the probability of spinning green

What is the probability of rolling an odd number?

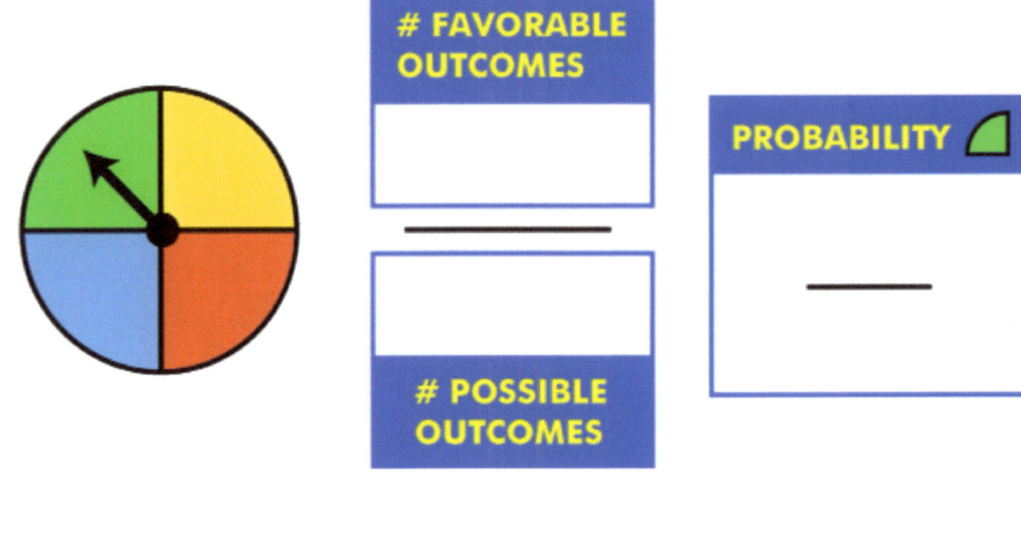

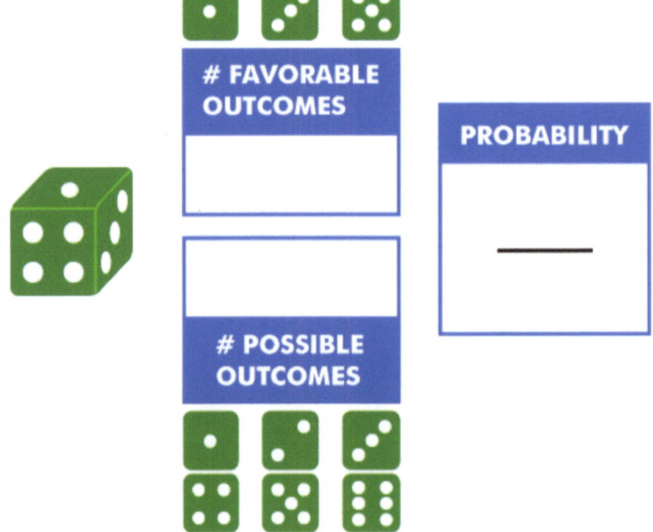

What is the probability of rolling a two?

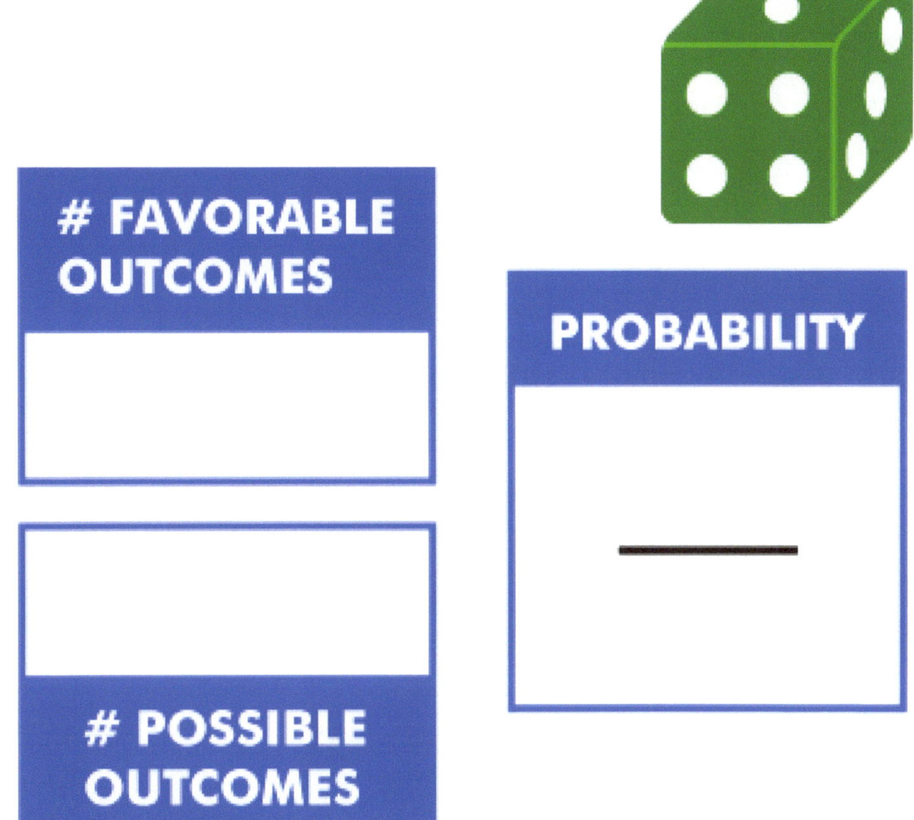

How many 2s do you think you would throw if you rolled a die twelve times?

Name_______________________________

Probability Quiz

1 True or false? Rolling a die and getting a 3 or a 5 are equally likely outcomes.

2 What is the theoretical probability of flipping a coin and getting a head?

1   2 $\frac{1}{2}$ $\frac{1}{4}$

A   **B**   **C**   **D**

3 What's the probability of spining blue?

4 What's the probability of spinning red?

Newburyport, MA 01950

1-800-596-3175

OnBoard Academics employs teachers to make lessons for teachers! We create and publish a wide range of aligned lessons in math, science and ELA for use on most EdTech devices including whiteboard, tablets, computers and pdfs for printing.

All of our lessons are aligned to the common core, the Next Generation Science Standards and all state standards.

If you like our products please visit our website for information on individual lessons, teachers licenses, building licenses, district licenses and subscriptions.

Thank you for using OnBoard Academic products.